The Evolution Sales System for High-Tech Businesses:

The One-of-a-Kind Turn-Key Global Solution with One Point of Contact. Take a Peek Inside our "Inner Workings" and Experience First Hand What You've Been Missing!

Angela Suzanne

TheEvolutionSS.com

This book includes 4 Customizable Sales Scripts to Boost Your Bottom Line, Engage Your Customers, and Always Leave Them Wanting More!

The Evolution Sales System for High-Tech Businesses: The One-of-a-Kind Turn-Key Global Solution with One Point of Contact. Take a Peek Inside our "Inner Workings" and Experience First Hand What You've Been Missing!

By Angela Suzanne

Copyright 2018 by Evolution Sales Publishing

All rights reserved. No part of this publication may be reproduced, distributed, or transmitted in any form or by any means, including photocopying, recording, or other electronic or mechanical methods, or by any information storage or retrieval system, without the prior written permission of the publisher and author, except in the case of brief quotations embodied in critical reviews and certain other non-commercial uses permitted by copyright law.

Book Cover Art by Evolution Sales Graphics

ISBN- 9781099156281

Printed in the USA

DEDICATION

To the other movers and shakers in the high-tech digital space who took me in and showed me the ropes. A special thanks to David, Gene, Joshua, Mark S., Mark A., Mike, Raymond, Ritchie, and Tony. This would not have been possible without all of you.

Introduction

Introduction

Hi! Welcome to the Evolution Sales System. The only Digital Worldwide Virtual Solution your Company will ever need moving forward. By taking a look inside the "inner workings" of this one-of-a-kind process you will see how easily you can transform your Live Events, Operations, and Sales into the well-oiled automated machine they were always meant to be.

Imagine never having to find highly qualified candidates again. Picture no longer managing day to day operations, training new staff, or overseeing your benefits, payroll, and healthcare departments. Would you even know what to do when all employee vacation and sick days have ZERO impact on your bottom line? Now, take one step farther with me and think of the peace of mind you'll feel once your business makes just as much money or even more whether you are in the office or attending the Super Bowl!

If you are ready to embark on this journey through the intricacies of The Evolution Sales System, then go ahead…
turn the page and come on in. Get ready to experience how fast this streamlined and

customizable process can be implemented. All you have to do is decide to turn the key…

Welcome and enjoy!

Chapter 1

Chapter 1

Have you ever had a client fall through the cracks? Lost an important business card? Or, misplaced the post-it or scrap of paper with a potential customer's contact information? If you've ever owned and operated a business, chances are you answered yes to ALL of these questions.

Although we know none of us would even have a business if it weren't for our customers, when did it become so difficult and all consuming to properly take care of them? When did the Customer Service Teams, Sales Teams, and Shipping Department start getting all of the attention? And how many more managers will you have to hire and train only to let them go months later, because things still aren't where they should be? Only to start the cycle over again and again with the same outcome?

Being a business owner can be so frustrating! There are a million reasons why most businesses don't make it. However, if you are reading this book then more than likely you not only love and appreciate your customers, but truly believe in the products and services you're offering them.

Moreover, finding solutions to all problems that arise is what you've done from day one. Plus, at the core you want what's best for all involved. So now let's simply connect the dots from you, to your staff, to your clients and back. Let's have your vision accurately represented, a staff who is as passionate about what you do as you are and customers who just cannot get enough and keep coming back for more!

The first step in the Evolution Sales System is consistently and effectively reaching out to your past, present, and future clients. In addition, you'll speak in their language, find out what they need and give it to them. Sounds easy enough, right? If your Company has ten customers or less, than sure! I have no doubt you could take care of all this from start to finish on your own.

However, the truth is if you're running a company generating over 2 Million Dollars a year than you have A LOT more than ten clients. Most of you are looking at over 100,000 at least! So, how the hell do you even begin pulling off a customized customer experience at this point? By walking through a specific example, you will see how it all comes down to extrapolating the right information from your customers quickly and efficiently, then taking this new knowledge and applying it in every other area of your business.

In this situation, there are two High-Tech Businesses both in the Survival Industry. Company A has an online store, Newsletter, Membership Site, and Live Events. Company B sells Camping and Survival Gear through their online store. Last year alone one company brought in around 2.4 Million and the other approximately 10 Million.

If you work and collaborate with any of these High-Tech businesses the list describing their Avatar or what their typical customer "looks" like will be provided right away. For survival companies the lists all look about the same and consist of the following:

1. Most clients are male and over the age of 60.

2. Many are former Police Officers or Military.
3. Most have some sort of religious backgrounds and believe in God.
4. Most are Republican.
5. If they are former Police Officers or Military they feel they are not as sharp as they once were. They want to stay relevant and current on what is going on and make sure they own the latest survival tools.
6. It's very important to them to have their wives, kids, and siblings protected and educated too.
7. About 50% are single, usually from divorce or being widowed.
8. At the Live Events 33% of the attendees are female.

To sum this up, if you are working with a High-Tech Survival Company the majority of their customers are retired men who are conservative, believe in God, and care about protecting themselves and their families. This means all of their marketing, advertising, and automated customer interactions prior to the Evolution Sales System have been based entirely off of this information alone.

Our goal is to change this altogether by connecting with the heart and soul of you, then completely transforming every interaction with your customers on all platforms moving forward. In addition, we provide your clients with an experience so unique and fulfilling they come back more frequently. And best of all, they feel YOU as the owner (with 100,000+) other clients still knows and cares about them on an individual level! All because you are now reaching out, informing them, and staying in contact in a way you never have before.

In the next Chapter you will see a specific example of how The Evolution Sales System can directly impact multiple areas of your Company from Marketing, Email Campaigns, Phone Scripts, and so much more!

CHAPTER 2

Chapter 2

How does a business owner with over 100,000 clients create both a customized and consistent start to finish customer experience? How many calls will need to be made? How many conversations will there be? How will you know when you've collected the right information?

Lucky for you the answer to ALL of these questions does exist! To demonstrate this and show you how it's done we'll use the two previous Survival Companies mentioned in Chapter One. Both Company A and Company B had customers who could be described with the Typical Survival Client List previously found on pages Nineteen- Twenty. Through talking with customers and staff at both Companies while using the proven Evolution Sales System our team uncovered the following:

Company A's clients consistently used the word "Hero" to describe themselves. On the other hand, we also learned Company B's clients considered themselves Heroes too. Like many other words in the English Language although spelled the same, the way in which they were used had two *very* distinctive meanings. And, in this case uncovering those differences had an enormous impact.

With Company A our team was talking with retired Police or Military men who still wanted to be a "Hero" to their families. More specifically, to their wives, daughters, and grandchildren. They found solace from knowing by purchasing certain survival tools and having their loved ones participate in specific defensive classes their family members would be safe and protected. Whether it was going for a run, attending college, or even taking care of errands, if heaven forbid, an attacker approached them they would have the necessary tools and knowledge to safely get out of the situation. By knowing their loved ones would now be safe while on their own, their Dad or Grandpa was able to feel like a "Hero" to those he cared about most.

On the other hand, Company B's clients also wanted to feel like "Heroes." However, to them, the same word, meant something entirely different. For Company B they had recently had an item with higher than normal returns. Initially, they had been unable to figure out why.

It turned out with their database, although they too were conservative men who believe in God and care about protecting their families, when they said "Hero" they actually meant "SuperHero." And yes, in this case we are talking about Superman, Batman, or even Captain America. Although, these men were retired they still desired the nostalgia and strength of their childhood heroes. When they purchased this particular item they wanted it to knock their opponent to the ground, not just disorient the attacker, thus enabling them to *feel* even for a brief moment as if they truly were their childhood "SuperHero."

Can you imagine how different the verbiage would now be for Company A versus Company B? How much will this now impact copywriting, advertising, and every future customer interaction? By specializing and customizing your client's experience each time they interact with your Company, this alone gives them more motivation to continue coming back again and again.

So how can you start getting this information out of your clients today? How do you begin this process? What's the first step to getting them to spill their guts? How does your company build a lifelong relationship with clients, sell more than ever before, and maintain the best customer experience they've ever had?

You now have two options. They both work. One will get you there much faster, but either will take you where you need to go. The choice is yours.

The first option, is to collaborate with Evolution Sales to run your Operations, Sales Teams and Live Events. This Turn-Key route gives you one stop shopping with ONE point of contact. In less than 30 days you can have a customized team working on your Companies behalf. Live Events, Sales Teams, Customer Service, Graphics, Logos, Click-Funnels, Membership Sites, Websites, and even Search Engine Optimization can all be managed, trained, and overseen.

Furthermore, you can pick and choose as many or as few of these departments as needed. All of which can be adjusted accordingly as your business continues to expand, evolve, and grow. In addition, Phone Scripts, New Products, and a unique customer experience will be created from start to finish specifically designed based on your needs and those of your clients. For example, Phone Scripts will be created through the proven Evolution Sales System. This means each Sales Representative has their own set of scripts which have been customized for the Company, the Owner, and them individually. This allows for your customers to have the most genuine interactions with your Company on every call moving forward.

To set up a FREE 30 Minute Phone Consultation and find out whether or not The Evolution Sales System is the right fit for your business simply email:

FreeConsultation@The EvolutionSS.com today!

Or, for those of you who prefer Option two, then keep reading...

CHAPTER 3

Chapter 3

By the end of the next two chapters you should never lose track of any customer again. Can you fathom what your life would look like already had you put this step-by-step system into place from day one? Or, what about starting today? Why not take all current clients and put them into the Evolution Sales System? How many fewer leads would you have to go out and find? This one step alone can save thousands of dollars to your bottom line every month.

Not only will you have all of your current and future clients in one place, you will also be able to access their information from any device anywhere in the world. As I write this book I am looking out at a beautiful mountain range from the top floor of my building. If it had been a week ago, my view was of an array of sailboats in the Pacific Ocean as I gazed out my window there.

In addition to having their contact information all in one place, you will also have their next step set up. Once it's entered, you don't have to think about it again. Why? Because, when it's time to reach out again this too will have already been set up.

Okay, so by now you're already asking who is Angela Suzanne and what is this Evolution Sales System anyway?

Let's go back to where it all started eighteen years ago. My first "real job" was working for a dot com Company called Talk2.com. Amazon and eBay were also just beginning. Everyone wanted to work at a "dot com" back then, and all of us hoped we had picked the right one, so when it went public we'd never have to work another day.

Back then, our Company was developing the technology needed to call in from your cell phone and listen to your emails being read back to you one by one. Eventually, it was to the point where you could also call in and hear current movie times in your area. Wow! It all seems so antiquated now, but at the time we thought we were on to the next best thing. Then, it had only been a few years since car phones and what seemed like brick sized cell phones were the latest and greatest tech gadgets! And let's not even get started on dial up internet connecting...

Now, we would all go crazy if we had to sit there on our phone while a computerized voice read us every last email in our inbox! Who has time for that?! Nonetheless, less than two decades ago it was mind blowing and a thrill to be a part of both the rise and fall of the dot com boom. Little did I know at the time, what a large role technology would be in everyone's personal and professional lives and what a big part it would play in the rest of my career. Fast forward a few years...

My Tuesday morning began like every other one had for the past few months. Each week our team was sent to a new state and every day a new city. The schedule was grueling to say the least, but I can't think of a better way to experience and see our beautiful country than to be interacting and training small business owners in almost all 50 States. It's fair to say when I woke up that morning, I had no idea how the rest of my life was about to be directed down an entirely new and challenging path.

At the time we had a three hour C.E. (Continuing Education) Accredited Technology Seminar consisting of training small business owners on how to organize and automate their clients. They also learned the importance of a website and SEO or Search Engine Optimization.

One day, about 30 minutes before the Live Event our main Speaker had a personal emergency and had to fly back home immediately. It was now up to me to step up and jump in. At the time, I was in my early 20's and most of the owners were 20-30 years older than me. To say I was nervous would be a complete understatement. Frightened would be much more accurate.

At the time, the expectation was to close 22% of the room into our $3,000 Twelve Week Technology and Training Program. We enrolled 75% of the room that day, and from then on I was given two assistants of my own and continued traveling as a Speaker/Trainer moving forward. During my time on the road many lessons were learned, but being able to engage an audience and close from the front of the room has proven to be invaluable in every business I've owned and client I have worked with since.

After teaching thousands of small business owners in over 40 States how to organize and automate their businesses I was able to do the same for all of my own projects moving forward. Back then, hardly anyone had their own website and SEO was an entirely new concept altogether. Nowadays, if you don't have your own website and aren't organically being pulled up in the first three options of a Google search you're not even considered relevant or taken seriously!

Since none of you really want to hear about every speaking engagement, sales job, and technology job I've ever had let's just skip right to what does matter. How do you organize and automate your business now, so you can immediately impact your bottom line, keep your customers thrilled, and have them consistently coming back again and again? And will implementing The Evolution Sales System work with your "type" of clients?

The answer is yes! So far this system has been tested in not only small to extra large Companies, but also a variety of industries. For example, just last year alone one client was a former CIA Agent who taught CIA centered Live Events for civilians, while another ran a Ten Million a year online camping gear company, and a third had an online organic skincare line. Each time the same online system was put into place and every time it worked!

So how do you get started? Why wait? Who doesn't want to spend less time working, have their business automated, and still make money even when they are on vacation with their friends or family? Then let's do it! In the next Chapter you will see how to start creating your own customized Client Engagement Scripts through The Evolution Sales System.

Chapter 4

Chapter 4

To begin effectively reaching out to customers, your Evolution Sales System proven Phone Sales Scripts will need to be created. This initial script is for your first voice to voice interaction. All scripts in the Evolution Sales System can be customized for each Sales Representative as well as align with your Companies' unique brand and vibe.

Script # 1- The Welcome Call

1. Introduction

2. State the purpose of your call.

3. Ask an open ended question(s).

4. Give them a coupon or offer code as a thank you.

5. Set up your next step.

Welcome Call Script Sample 1

Introduction:

Hi _____. This is _____ calling on behalf of _____.

How are you today?

Purpose of the call:

(Owner's name here) really wants to go the extra mile this year, so we are reaching out to as many clients as possible.

We want to get feedback from you to find out what we are doing well and what we could improve on to make your experience even better.

Next, you want them to say yes or agree with you early on.

This question is the one I use the most:

Does that sound okay?

Another great option is simply:

Right?

Open ended question:

To get us started, what is your favorite product?

Or...

To get us started, what is your favorite thing about (Owner of the Company?)

Then, sit back and let them talk! Take a lot of notes, then after the call you will enter these into your already created Client Google doc or current Contact Management System.

Coupon or Offer Code:

At the end of your call always offer them a discount or offer code for their time. Anything from a free digital download to 10% off their next purchase works great. Here is what this looks like in a script form:

Thank you _____ for your feedback. (Owner's Name) will be happy to hear what you had to say. As a thank you for taking the time to speak with me today I have a FREE Digital Download for you to use within the next 24 hours. Do you have a pen to write this down or would you prefer I email it to you?

Next step:

Then, to wrap up the call make sure there is a next step. It's an absolute must to let them know when you will be following up and why.

It looks like you will be due to order another _____ in about 2 weeks. Would it be okay if I checked in on you then?

Welcome Call Script Sample 2

Introduction:

Hi _____. This is _____ calling on behalf of _____.

How are you today?

Purpose:

We call everyone within 24 hours to welcome them to _____. The purpose of this call is to introduce myself. My name is _____ and I will be your direct contact for anything you'll ever need moving forward.

Open ended question or statement:

Tell me about yourself.

Coupon or offer code:

As a thank you for being one of our loyal customers I have a 10% off coupon code for you to use within the next 24 hours. Do you have a pen or would you prefer I email it to you?

Would it be okay if I followed up with you in 2 weeks?

Having a phone script is necessary, but using a customized script is better. Here is one quick part you can customize for yourself and/or your Sales Representatives. First, pay attention to how you would normally greet someone when you call them. How do you start a conversation with a friend, colleague, or even your Mom?

Do you open your calls by saying:

- Hi

- Hello

- Good morning

This may seem simple, but it does work. The more you can make this natural, the better. If you are reading a script, people can tell. You seem less sincere, less engaging, and most importantly are less likely to get a warm response from them.

On the other hand, if your Sales Representatives put more of *them* into your scripts they'll now come across as natural, real, and caring. People will feel the pulse of them and your brand all while being more willing to hear what your Company has to offer. It breaks down barriers faster and gets your foot in the door!

To review, when using the Evolution Sales System Welcome Call Script make sure to:

Script # 1- The Welcome Call

1. Open with a natural greeting for your introduction.

2. State the purpose of your call.

3. Ask an open ended question(s).

4. Give them a coupon or offer code as a thank you.

5. Set up your next step.

Congratulations! Now that you have the first call down, let's move on. The only script you'll ever need to set an appointment where your Sales Representatives can build credibility and pre-qualify clients for your Top Level Products, Services, or Live Events!

Chapter 5

Chapter 5

Have you ever noticed how little you hear from your most satisfied customers? You know the ones I am talking about. Those who enjoy the products you are selling, rave about your Live Events, and cannot get enough of anything and everything your Company is offering?

Then on the other hand, have you ever had someone who is just NOT the right fit for you, your products, or your services? They don't like them, something is always wrong, they don't answer when you call to try to fix it, and no matter what you do or how far backward you attempt to bend it's still never enough?! How nice would it be if there was a proven system to find out more about potential clients and discover *ahead* of time if they truly are the right fit for you and your brand? Imagine, actually being able to predict whether they will stick or split?

Well, my friend this system does exist. Furthermore, it's easier than you think. Plus, it's effective, efficient and will save you boat loads of time, energy, and effort along the way. This leads us to our Second Phone Sales Script.

This Script consists of setting an appointment for the Live Event/Product Expert to get to know the customer and see if they are the right fit for your most expensive product or service. For this example, we will be using a Live Event, but you can interchange Live Event for any product or service your Company offers. This script is meant to keep your clients informed and engaged. If it's not a win/win for both of you then it's better to find out now. By the end of this phone appointment:

1. Find out more about them.

2. Inform them more about your products, Company, and/or Live Events.

3. Pre-qualify them for your high-end product or service.

4. Set up their next step for a specific date and time usually within 24 hours to meet with the Closing Sales Representative.

Script #2 - Updates and Announcements

Hi _____!

We are calling to announce our most recent Live Event coming up October 15th. We are reaching out to our most valued customers first as there are limited seats available.

Have you ever thought about attending one of our Live Events?

If they respond with yes, then say:

Great! What is it about attending one of our Live Events that interests you most?

Then let them talk. Listen to what they say. Take notes. From there you'll repeat back what they said and tie in how this new Live Event would include one to two of the items they mentioned and then add one to two more for added value.

It may look like this:

Customer: What interests me most is being around other like minded individuals. By being there in person it would motivate me to really learn as much as possible. And, it would be nice to meet others who are also going through the same process I am and have the opportunity to help each other grow and succeed.

Representative: I couldn't agree more with the value of being surrounded by others like you who are also going through the grind of growing their businesses. Due to feedback we've received from customers like you, you'll be happy to hear we are adding a NEW class this session primarily focusing on _____. How does that sound?

Customer: That sounds great!

Representative: Good. How would you like to set up a 30 minute phone appointment with one of our Live Event Experts (Closing Sales Representatives) to find out more specific details about this new Live Event and to see whether or not this is the right fit for you?

Customer: Yes. That works.

Representative: Perfect. Let's get you scheduled.

To review, when using the Evolution Sales System's Updates and Announcements Script make sure to:

1. Always open with a natural greeting/Introduction you would normally use.

2. State the purpose of your call. In this case to announce a new product or Live Event.

3. Find out their interest level regarding the new product or Live Event.

4. Repeat and add value to their comments.

5. Set up their next step. In the above example, simply ask to set up a 30 minute phone appointment with the Closing Sales Representative.

Chapter 6

Chapter 6

How many of us have been in the middle of dinner and the phone rings. You don't recognize the number, so you answer it just in case, only to find out it's another telemarketer. Ugh!! And have you EVER answered and then wanted to stay on the phone longer to fill them in on your entire life story? I am sure it is more than safe to say this has NEVER once happened.

Since we all know how it feels to receive an unwanted call, then of course this is the last thing we want our customers to ever experience when we reach out to them. So the question is, how do we not only get them to answer our calls, but welcome them, and on top of it all still have them spill their guts to us? The answer is by using the next script in the Evolution Sales System.

This script consists of getting your clients to spill their guts in 15 minutes or less. Yes we have timed it, yes it works, and yes the client is talking the majority of the time.

Script #3- Qualifying the Client

1. Always open with a greeting you would normally use.

2. State the purpose of your call. In this case to get to know more about them, and to have them find out more about the Company to see whether or not it's a good fit for both.

3. Listen. Let them tell you as much as possible.

4. Recap what they said.

5. Ask them if there is anything you missed.

6. Set up their next appointment.

Script #3 - Qualifying the Client

Introduction:

Representative: Hi _____! This is _____ calling for our 2:30 appointment. How are you today?

Customer: I'm alright. You?

Representative: I'm good. Thank you for asking.

Purpose of the call:

The purpose of this call is for us to find out a little more about you and for you to find out a lot more about our new Live Event and/or product. We have found 100% of the time by the end of this call we'll both know whether or not it's the right fit.

Does that sound okay?

Customer: Sure!

Open Ended Question or Statement:

Representative: Great. To get us started why don't you start by telling me a little more about yourself.

Now... pay close attention to this. It's the hardest part. Ready? You have to be quiet and LISTEN. That's it! Don't be afraid of the silence. It's the best kept secret. Silence is our friend! It's everything.

The truth is our clients don't like silence any more than we do. So, ask them to tell you about themselves then stop talking and wait for them to respond. At times it might seem like entire minutes are passing. I assure you they are not.

If you are really struggling with the silence then count to ten slowly. By the time you are done they will be spilling their guts. Before you know it, they'll go on and on and on.

Maybe this sounds too good to be true. It's not. Last year alone we had multiple clients in a variety of industries we tested it out on. Here are actual client discoveries collected using the Evolution Sales System. The Sales Representatives took notes during their calls, then typed these up right after while the conversations were still fresh. Names and locations have been changed to protect both Company and client information, but the quality received is all there.

Client Discovery # 1 - Brandon Hill

Brandon is turning 64 in February. He is divorced, but still best friends with his former wife. He has 4 grandkids who he teaches piano to every Saturday. They are his daughter Summer's kids. He grew up in a 5th generation religious household in Portland, Oregon with 8 brothers and 7 sisters. He also went on a Church Mission.

Although, he no longer considers himself an active member of that particular religion, he was surrounded by learning about the "end of days" through studying Revelations in the Bible throughout his childhood. He feels this is something he has been told his entire life.

His brother has always been way into the Survival Industry and teaches as a Professor at the State University. His brother has also led multiple 28 day survival hikes. This brother is all about living "off the grid." This same brother has had and continues to have a huge influence on Brandon and his views on why survival is necessary and relevant.

Although he does not live his life in fear, he is subscribed to our weekly newsletter and cannot get enough of it. He called his former wife just this morning to tell her all about the latest one. He also talks to his kids and siblings. He wants to start a weekly family meeting to discuss the information found in these newsletters and make sure everyone he loves is prepared and safe.

This year his main focus in on survival. He wants to start gardening, learn how to build his own guns, and learn anything and everything that is available to him in this category. He then plans on sharing the information he attains through these weekly meetings with his family.

Brandon went on to attend a Live Event a few months later and is still an avid fan and customer for this Survival Company.

Client Discovery # 2 - Richard English

Richard is 60 years old. He grew up in Anaheim and Long Beach, California where he attended a Christian School. In addition, he was a former fighter pilot in the Navy for 15 years. His son Anthony is currently in the Navy and lives in Oak Harbor, Washington. He is a single Dad, which can be challenging, but he wouldn't change a thing.

Richard has an oldest son who he didn't say much about except he is the difficult one. His third son Ben just returned from Quantico where he was hoping to begin his career as an FBI Agent. This unfortunately did not work out as planned. They are clearly a huge military/government family.

Richard has not attended any of our Live Events. He has read the owner's book and thinks the world of him. He and the owner are definitely on the same page with what's going on in the world.

Client Discovery #3 - Trevor Hill

Trevor is an engineer in his 60's. His wife is a homemaker. He carries around the companies most popular product and has it in his pocket as we speak. He just read about a new defensive move this morning in the weekly newsletter. He already sent an email to his two daughters and will tell his wife about it tonight. Trevor wants his family to be as prepared as possible.

Client Discovery #4 Emily Carlton

"Sam (the owner) does his job very well. He can take the world and apply it down to the everyday person. He also has great stories. Loved his stories. You wish you could just have a day of stories. I feel like I am adding layers to my life. Sam is like my hero."

Emily is a Mother of two. Her son attended one of Sam's Live Events with her. Her kids just went to college in Boston and need to be able to take care and defend themselves if needed. She was very impressed I called personally to reach out.

Client Discovery #5 Kathy Russell

She is an author. She writes Fiction and compared it to the Twilight Series. She is currently working on her third book. Her series is called Ocean Depths. She has won multiple awards. Mom's choice Award, etc. She and her husband have two homes. One in Texas and the other in Florida.

She and her husband are gun owners and consider themselves very outdoorsy. They love shooting together which is how they got interested in Sam and what he has to offer.

They also have Sam's book and watch all of his videos. She said, "I am no longer afraid."

Client Discovery #6 Taylor Lewis

He is a current weekly newsletter subscriber. Taylor has also ordered multiple other items. He was wearing his gun belt for the first time today and said, "it seemed like high quality." He is also enjoying the informative emails he receives and said they have good information.

Client Discovery #7 Steve Mitchell

Steve likes the newsletter and finds it both applicable and helpful. He signed up for it in the first place, because he thought it would be interesting and useful and it is!

Client Discovery #8 Kyle Smith

He really liked the first day of instruction at the 2 Day Live Event he attended in Washington, DC. He specifically enjoyed the lock picking information. He also found the life story examples very engaging.

Client Discovery #9 Thomas Money

Goes by Tom. He is a huge fan of the owner. He just opened up a custom gun holster business. His wife Vienna, daughter Mercedes, and his son-in-law are all involved. He has gone through all of the escape techniques previously from his Military Training. He is also currently an ER Nurse at the VA hospital.

He would like his daughter and wife to know the information. "Family is not blood, it's so much more."

Client Discovery #10 Jody Aguirre

Jody lives with her husband in Erie, Pennsylvania. She is 72 years old and loves gardening. Her husband is retired, but used to teach college. They are about to celebrate their 50th Wedding Anniversary in Orion, Ohio where they are both from. She said her Mother is especially proud of her, because she is the only sibling to reach this milestone.

There are so many of the products she uses and she thinks they are wonderful. She couldn't pick her favorite, because she uses so many and likes them all.

She and her husband are also heading to Orlando for their Nephew's wedding. We set up a follow up call to check back in the first week of April after all of her trips. She was very happy we were here to help and could already see how much this could benefit her.

She definitely loves the products and just wants to see a more user friendly way of ordering.

Client Discovery #10 Caleb Devereaux

Caleb was enrolled in one of our previous Live Events. His wife came on the trip with him and was supposed to shop while he went to class. When they arrived at the hotel their room wasn't ready, so she came to the first couple of hours and liked it so much they ended up buying her a ticket to stay for the entire time! These two love the owner and were glad they took the class.

As you can see from the above examples, not only are you getting to know your clients in a way that's never been done before, but you are uncovering so many additional gems as well. There are testimonials you could use with permission of course on your newsletters, websites and email campaigns. If enough people mention the same thing you can source it and look into selling it on your site. Then, when it comes out tell everyone this product was launched due to direct feedback from them. Clients soak this stuff up! It's now a win/win for all of you.

And last, but certainly not least is the one thing that has been overlooked thus far by every client I have ever worked with from day one. Through the discovery/qualifying call at the core your Sales Representatives are really just getting to know people inside and out. By doing so they'll now know which products, services, or Live Events are the right fit for them. Once rapport and trust are built the sky really is the limit. To this day if I call a former or current client and tell them something they need to do or buy they do it 100% of the time. Why? They know I am only recommending it, because I know they'll love it. The relationship has been properly and effectively built over time. The trust is there.

Another element we uncovered again and again is most clients are there, because they are already a fan. By making them feel more involved in the Company they really do get engaged and keep coming back again and again. Everyone likes to feel like a part of the "in crowd."

To review, when using the Evolution Sales System Qualifying the Client Sales Script make sure to:

1. Always open with a greeting you would normally use.

2. State the purpose of your call. In this case to get to know more about them, and to have them find out more about the Company to see whether or not it's a good fit for both.

3. Listen. Let them tell you as much as possible.

4. Recap what they said.

5. Ask them if there is anything you missed.

6. Set up their next appointment.

CHAPTER 7

Chapter 7

Being an effective closer takes everything you've learned so far and ties it altogether. At the end of the day if you can't close your customers and take their payments, then you won't stay in business. A deal is NEVER considered "closed" until it's signed and their payment has been processed. Period.

This script is vital to utilize correctly. Your mindset during the closing call needs to be calm. In addition, make sure you are actively listening to everything they say and take detailed notes on any concerns and objections they have. The remainder of your call will consist of overcoming their objections one at a time. Then, the other crucial part of closing is making sure the client says out loud WHY they want this product or service and how it will positively impact their life. They need to hear themselves say it. In turn, they literally end up closing themselves! It's beautiful to see.

Script #4- The Closing Script

1. Always open with a greeting you would normally use.

2. Listen to any concerns and objections they have.

3. Overcome their concerns and objections one by one.

4. Make sure they state why they would be a good fit for the program and how it would benefit them.

5. Take their payment and congratulate them on enrolling!

Chapter 8

Chapter 8

Way to go! By now you should have 4 Customized Scripts for your Company. Once The Evolution Sales System is properly implemented each Phone Script will not only be customized for your Company, but tailored to your individual Sales Representatives too. Your Company now has its own proven system to reach out and engage your current and new customers in a way none of your competitors are currently doing.

Along the way new fresh testimonials will be uncovered, invaluable feedback will be unearthed, and most of all more clients will buy the right products and services. Plus, your Sales Representatives will have a system that works and clients will feel individually valued and cared for as your bottom line increases.

Thank you for going on this journey with me. Remember, if you prefer the "Turn-Key" approach simply email:

FreeConsultation@TheEvolutionSS.com

to set up an appointment directly with one of our highly qualified Representatives within the next

24 hours. Make sure to mention this book and receive 50% off of your on-boarding fee, which is worth up to $10,000!

All my best to you and your business.

About the Author

Angela Suzanne has traveled all over the Nation working with over 1,000 Owners of High-Tech companies as well as Small Businesses in more than 40 States. Through her customized consulting, presentations and trainings she has collaborated with Universities, Entrepreneurs, Worldwide Companies, and even Mom and Pop
shops. Through enhancing the customer experience, creating customized Phone Scripts,

developing Top Level Products and Creating One-of-a-Kind Global Live Events both customer engagement and satisfaction increase while propelling revenue to the next level.

These travels have led to appearances on HBO, TNT, ESPN, and even the Disney Channel. Although Angela has generated over Three Million Dollars for the University of Phoenix and Two Million Dollars for The Coaching Institute, nothing compares to reaching over One Million dollars for Jerry's Kids, ALS, The March of Dimes, Make a Wish, Primary Children's and Shriner's Hospitals.

If your Company is currently making between 2-10 Million Dollars annually and you are looking to reach your next milestone email:

FreeConsultation@TheEvolutionSS.com today!

www.ingramcontent.com/pod-product-compliance
Lightning Source LLC
Chambersburg PA
CBHW072222170526
45158CB00002BA/703